The SOLAR SYSTEM

Robin Kerrod

Lerner Publications Company • Minneapolis

This edition published in 2000

Lerner Publications Company
A Division of Lerner Publishing Group
241 First Avenue North, Minneapolis MN 55401

Website address: www.lernerbooks.com

© 2000 by Graham Beehag Books

Library of Congress Cataloging-in-Publication Data

Kerrod, Robin
    The solar system / Robin Kerrod.
        p. cm. – (Planet library)
    Includes index.
    Summary: Introduces the solar system, its planets,
moons, asteroids, and comets, and its exploration.
    ISBN 0-8225-3903-9 (lib. bdg.)
    1. Solar system—Juvenile literature. 2. astronomy—
Juvenile literature. [1. Solar system. 2.Astronomy.]
    I. Title. II Series: Kerrod, Robin. Planet library.
    QB501.K52   2000   98-34708
    523.3—dc21

Printed in Singapore by Tat Wei Printing Packaging Pte Ltd
Bound in the United States of America
1 2 3 4 5 6 – OS – 05 04 03 02 01 00

CONTENTS

The planet Saturn with a few of its many moons. Saturn's shining rings make it one of the most beautiful members of the Sun's family.

Introducing the Solar System

Earth, the planet we live on, belongs to a family of bodies that travel together through space. Earth is part of the Sun's family, which is called the solar system. The word *solar* means "having to do with the Sun." The Sun is a star—a huge ball of glowing gas.

Earth is one of nine large bodies that circle in space around the Sun. We call these bodies the planets. Many other bodies belong to the solar system. Smaller bodies known as moons circle around most of the planets. Smaller still are swarms of rocky lumps we call asteroids. They circle in a huge ring between the planets Mars and Jupiter. In addition, smaller particles known as meteoroids hurtle through space around the Sun.

Some of the most fascinating and least-known members of the solar system are the dusty "snowballs" we call comets. They visit our skies from time to time, often growing spectacular tails.

The newest members of the solar system are spacecraft that we have launched into space. These include satellites that circle Earth, and space probes that journey to the Moon and to distant planets, asteroids, and comets.

It is thanks to these spacecraft that we know so much about our solar system. These craft have also shown us that solar systems exist around other stars in the heavens. So there are probably many other planets in our Universe. Some of these other planets could possibly be like our home planet, Earth.

In the Beginning

The solar system was born from a huge cloud of gas and dust. The Sun, the planets, and other bodies formed when the cloud collapsed under gravity.

Long before the solar system was formed, there was nothing in our corner of space but a great cloud of gas and dust. Many clouds like this exist throughout space. Astronomers call them nebulae. It is in nebulae that stars are born. The nebula that once occupied our corner of space was the birthplace of the star we know as the Sun.

This nebula started to shrink, or collapse, about 4.6 billion years ago. The collapse happened because of gravity. Gravity is the pull, or attraction, that every bit of matter has on objects on or near it. Specks of dust and wisps of gas in the nebula began to attract one another and to form a denser, or more tightly packed, mass.

SPINNING AROUND

As the cloud started to shrink, it also started to spin around. Over time, it turned into a thick disk with a large bulge in the center. At this stage, it would have looked something like a fat spinning Frisbee.

The bulge at the center kept on shrinking. As it became denser, it began to heat up. The heat came from the energy in the colliding particles. Now looking more like a ball, the bulge got smaller and smaller, and hotter and hotter.

The disk of matter around the hot ball gradually flattened out. Small lumps of rock and metal filled the warm inner parts of the disk. In the cold outer parts, there were lumps of ice and freezing gases.

The time is nearly 5 billion years ago. A huge cloud of gas and dust in space begins shrinking (1) and starts to spin around (2). As it shrinks more and more, it starts to flatten out (3). Inside, it is warming up.

1

Birthplace of Stars

Stars galore are being born in nebulae all the time. From Earth, we can easily see one of these nebulae with the naked eye in the constellation, or star pattern, we call Orion. This constellation stands out in the night sky in winter. The Orion nebula lies near the line of three stars in the middle of the constellation. The nebula glows brightly because it reflects light from young hot stars within it.

The Orion nebula, photographed by the Hubble Space Telescope

2

3

THE SUN SHINES

The hot ball in the center of the spinning disk continued to shrink until it was about 900,000 miles (1,400,000 km) across. By this time, it had become extremely hot. Inside, its temperature rose as high as 27,000,000° F (15,000,000° C).

At this temperature, atoms of hydrogen gas in the ball began to combine, or fuse together. Fantastic amounts of energy were produced in this process. The ball gave off this energy as light and heat. It began shining as a star— the star we call the Sun.

THE PLANETS FORM

Little lumps of material were whizzing around in the spinning disk surrounding the new Sun. They were continually bumping into one another and sticking together to form larger and larger lumps. Gravity was again at work. Over millions of years, the lumps grew and eventually formed the planets and their moons.

Four planets—Mercury, Venus, Earth, and Mars—formed in the warmer, inner part of the disk. They formed from the heavier lumps of rock and metal found there. Five planets formed in the colder, outer part of the disk— Jupiter, Saturn, Uranus, Neptune, and Pluto. They formed from the lumps of ice and freezing gas found there.

Above: The planet Mars, photographed by the Hubble Space Telescope. It is one of the four rocky planets, which formed in the warmer, inner regions of the solar system.

5

4

The rotating cloud of gas and dust continues to flatten into a disk (4). At its center, a ball of matter grows, heats up, and starts to glow. Around it, bits of matter start to lump together (5).

Over time, the ball in the center
starts shining as the Sun, and
the circling lumps turn into
planets. The solar system as we
know it has been formed (6).

6

Below: The planet Neptune,
photographed by the Hubble
Space Telescope. It is one of
the four giant planets, made up
mainly of gases, which formed
in the colder, outer regions of
the solar system.

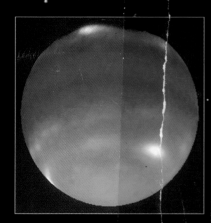

THE LEFTOVERS

Many smaller lumps of matter were left over after the
planets and their moons had formed. They were
scattered throughout the space between the planets.
Some of these lumps gathered together in the space
between Mars and Jupiter to form a huge ring of material.
We call them the asteroids.

Some lumps, known as meteoroids, rained down on the
newly formed planets. They dug out deep holes, or
craters, in the planets' surfaces. Meteoroids still circle the
Sun. Other lumps of matter remained on the edges of the
solar system. We see them only when they wander in
toward the Sun and start to shine as comets.

IN THE END

The solar system will probably stay much the same as it
is for another 5 billion years. Then the Sun will start to
die. It will probably grow bigger and bigger until it
stretches out as far as the planet Venus. The Earth will be
baked by the Sun's heat. The solar system as we know it
will come to an end. Astronomers predict this ending for
the solar system because they have seen the same thing
happen to other stars like the Sun.

Mapping the Solar System

The solar system stretches over a vast region of space. In a space shuttle, it would take you more than 100 years to travel from one side to the other.

The Sun lies at the heart of the solar system. The nine planets that circle around it form the main part of the Sun's family. They circle around the Sun at different distances. The closest planet to the Sun is Mercury. Then, in order going out from the Sun, are Venus, Earth, Mars, Jupiter, Saturn, Uranus, Neptune, and Pluto.

PLANET ORBITS

Each planet travels around the Sun in a path we call its orbit. The orbit never changes. Most of the planets travel in orbits that are nearly circles. Others travel in orbits that are more oval, or elliptical, in shape. Mercury and Pluto have very elliptical orbits. The Sun keeps the planets in their orbits with its enormous gravity. This pull is so powerful that it acts over distances of many billions of miles.

Below: The orbits of the four inner planets—Mercury, Venus, Earth, and Mars

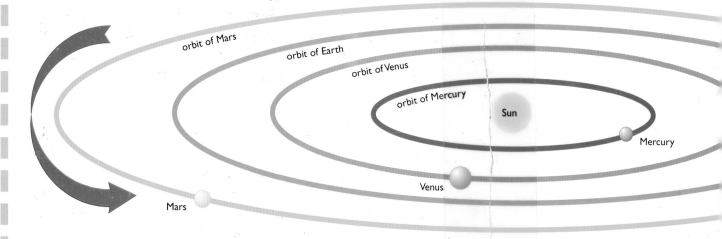

orbit of Mars

orbit of Earth

orbit of Venus

orbit of Mercury

Sun

Mercury

Venus

Mars

STAR POINT

Without the Sun's gravity to hold them in orbit, the planets would fly off into space.

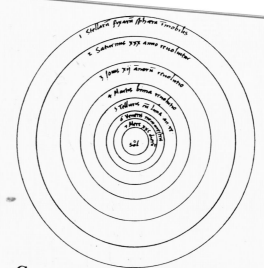

Copernicus's Brainwave

Until a few hundred years ago, everyone thought that Earth was the center of the Universe. This idea was set down by an ancient Greek astronomer named Ptolemy. But in 1543, a Polish astronomer named Copernicus disagreed. He said that the Sun was the center of the Universe, and that Earth was simply a planet that circled around the Sun. The work of later astronomers such as Kepler and Galileo proved that Copernicus was right.

Mercury's average distance from the Sun is about 36 million miles (58 million km). At its farthest distance, Pluto travels more than 5 billion miles (7 million km) away. Comets and other small bodies lie at even greater distances. All the bodies in the solar system are far apart. In fact, most of the solar system consists of empty space.

THE INNER PLANETS

The four planets nearest the Sun, from Mercury to Mars, lie close together compared with the other planets. They form a small family of their own—the inner planets. Earth is the largest of the inner planets. The other three inner planets are also rocky bodies like Earth. They are often called the terrestrial, or Earth-like, planets.

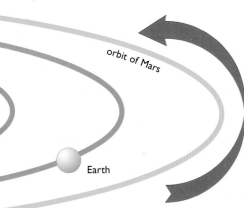

orbit of Mars

Earth

Astronomical Units

The solar system is very large. From one side to the other it measures more than 10 billion miles (16 billion km). It is difficult for anyone to imagine distances so large. So astronomers often measure distances in astronomical units (AU). One astronomical unit is the distance between the Sun and Earth, about 93 million miles (150 million km). Using this unit, distances become easier to understand. Neptune, for example, is 30 times farther away from the Sun than Earth is.

Planet	Distance from Sun (AU)	Planet	Distance from Sun (AU)
Mercury	0.4	Saturn	9.5
Venus	0.7	Uranus	19.0
Earth	1.0	Neptune	30.0
Mars	1.5	Pluto	39.0
Jupiter	5.2		

THE OUTER PLANETS

The inner planets make up only a tiny part of the solar system, as the drawing at the bottom of the page shows. The remaining planets—Jupiter, Saturn, Uranus, Neptune, and Pluto—take up a far larger part of the solar system. We call them the outer planets. They lie much farther from the Sun, and from one another, than the inner planets do.

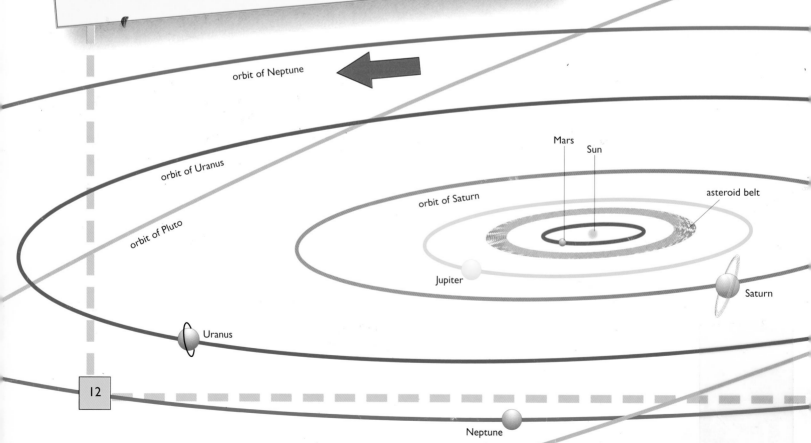

orbit of Neptune

orbit of Uranus

orbit of Pluto

orbit of Saturn

Mars
Sun
asteroid belt

Jupiter

Saturn

Uranus

Neptune

A beam of light from the Sun takes less than 9 minutes to reach Earth, but more than 5 hours to reach Pluto.

SAME PLANE

The drawing below shows another feature of the solar system. Most of the planets travel in the same plane. This means that if you could place the Sun and Earth on a flat sheet, the other planets would be on the sheet too. The odd planet out is Pluto. Pluto travels in an orbit that takes it far above and below the plane in which the other planets travel.

SAME DIRECTION

All the planets also travel in the same direction in their orbits around the Sun. If we could look down on the solar system from a point in space high above Earth's North Pole, we would see the planets traveling counterclockwise.

BEYOND THE PLANETS

Pluto is the most distant of the planets. But it is not the most distant body in the solar system. Much farther out are great clouds of small icy bodies. These clouds may reach out trillions of miles from the Sun. From time to time, some of the icy lumps leave the clouds and journey in toward the Sun. We then see them shine as comets.

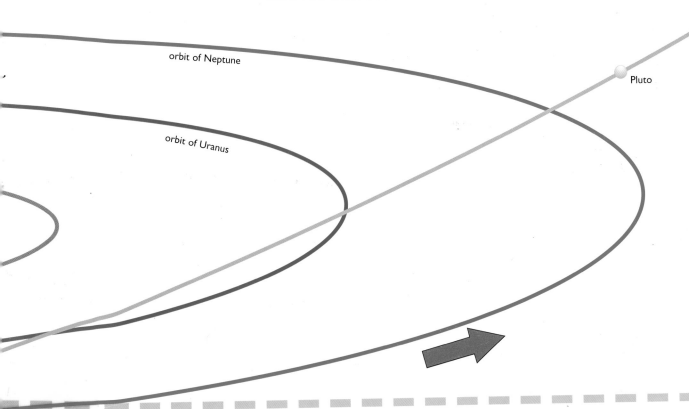

orbit of Pluto

orbit of Neptune

Pluto

orbit of Uranus

Family Portraits

The many members of the Sun's family are all very different from one another. They vary in size, makeup, and appearance.

STAR OF THE SOLAR SYSTEM

The Sun is quite a different body from all the other members of its family. It is a huge ball of white-hot gases, so big that it could swallow a million bodies the size of Earth. It is a kind of body we call a star. It looks so much bigger and brighter than the other stars in the sky because it is much closer to us.

The Sun is the only body in the solar system that gives off light. The Moon and the planets shine in the night sky, but only because they reflect light from the Sun. If the Sun did not shine, the whole solar system would be in darkness.

This close-up of the Sun's surface shows a boiling, bubbling mass of very hot glowing gas.

COLD PLANETS

There are four other huge balls of gas in the solar system. They are four of the outer planets—Jupiter, Saturn, Uranus, and Neptune. We call them the gas giants. But they are not hot like the Sun. They are very cold and are made up mainly of gas and liquid gas. The other outer planet, tiny Pluto, is very cold too.

... AND HOT PLANETS

The four warmer, inner planets—Mercury, Venus, Earth, and Mars—are much smaller and are made up mainly of rocks. Our home planet, Earth, is the biggest rocky planet. Its rock is cold only near the surface. We see hot rock from below when it forces its way up to the surface in volcanoes.

Above: Saturn is one of the cold gas giants of the outer solar system. At the top of its atmosphere, the temperature is about –300° F (–185° C).

Below: Molten rock, or lava, pours out of a volcano on Earth, one of the small hot planets of the inner solar system.

Saturn has 18 known moons —more than any other planet. The largest ones are shown in the picture. In the foreground is Dione. At top right is Saturn's largest moon, Titan, which is bigger than the planets Mercury and Pluto.

This is the minor planet, or asteroid, we call Gaspra, a rocky body only about 10 miles (17 km) long. The space probe *Galileo* took this photograph when it flew past the asteroid in 1991.

MANY MOONS

Smaller than the planets are the bodies known as moons, which circle most of the planets. Only Mercury and Venus do not have moons. Earth has one moon, which we call the Moon. Mars has two moons. Between them, the four giant planets have at least 59 moons. The two biggest moons in the solar system are Jupiter's Ganymede and Saturn's Titan. They are both bigger than the planets Mercury and Pluto.

Moons are made up of different materials. Our own Moon is made mainly of rock, much like Earth. The large moons of the giant planets are a mixture of rock and ice. Jupiter's moon Io is unusual because it has volcanoes erupting all over it. No other moon has volcanoes.

MINOR PLANETS

Between the orbits of Mars and Jupiter is a ring of small rocky bodies called asteroids. They are also known as the minor planets. Even the biggest, called Ceres, is only about 600 miles (1,000 km) across. Astronomers once thought that these bodies were the remains of another planet that broke up into pieces long ago.

Above: Thousands of years ago, a meteorite landed in Arizona and created the famous Barringer Crater.

FALLING STARS

Sometimes in the night sky you can see little streaks of light. It looks as if some of the stars are falling down. The stars are not falling, of course. What you see are meteors. These are fiery streaks made by specks of rock called meteoroids. These specks hit Earth's atmosphere traveling very fast. Friction from the air heats the meteoroids. They glow red-hot and then burn up, leaving behind a fiery trail.

Some of the bigger meteoroids do not burn up completely. They fall to planets and moons as meteorites. Large ones dig out deep pits, or craters, in the surface.

Below: The best known of all "hairy stars," or comets, is Halley's comet. It last appeared in our skies in 1986 and will not be seen again until the year 2061.

HAIRY STARS

Comets are some of the smallest members of the Sun's family but are among the most spectacular to look at in the night sky. Just a few miles across, these icy lumps visit our skies after traveling for billions of miles from the outer parts of the solar system.

As they near the Sun, comets give off great shining clouds of gas and dust. These clouds often fan out from the comet to form a tail millions of miles long. Because of these tails, some ancient astronomers called comets hairy stars.

Wandering Stars

The planets shine in the night sky like bright stars. But they change their position night by night against the background of true stars.

If you look at the night sky at about the same time for several nights, you will see the same patterns, or constellations, of stars. Ancient astronomers called the stars in the constellations the fixed stars. They called the planets wandering stars. The planets looked like stars but were constantly on the move among the never-changing constellations. The word *planet* means "wanderer."

You can see another difference between planets and stars. Stars twinkle, but planets do not. Air currents in the atmosphere make the faint light from the distant stars wobble. This causes them to twinkle. The stronger light from the planets is not affected, so they shine steadily.

BRIGHT AND BEAUTIFUL

From Earth we can see five planets without a telescope. Venus is the brightest by far. It shines brighter than any star. Both Venus and Mercury can sometimes be seen just before sunset or just after sunrise. Mars, Jupiter, and Saturn also

In the night sky, the planets can be found within an imaginary band called the zodiac. The zodiac contains several star patterns, or constellations. Ancient astronomers named the constellations after figures they imagined they could see in the patterns of stars. We still use these names for the constellations of the zodiac.

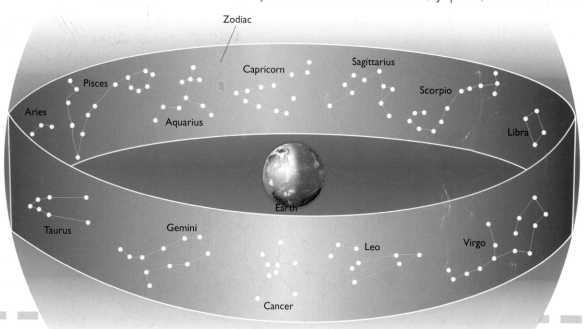

Zodiac

Pisces
Aries
Aquarius
Capricorn
Sagittarius
Scorpio
Libra
Taurus
Gemini
Earth
Leo
Virgo
Cancer

shine brightly in the night sky. They are easy to tell apart. Mars shines with a reddish-orange color. It is often called the Red Planet. Jupiter appears to be a brilliant white, and Saturn's light is a yellowish white.

THE CIRCLE OF ANIMALS

The planets do not wander everywhere in the night sky. They can be found only among certain stars. The band of stars in which they are found is called the zodiac. The word *zodiac* means "circle of animals." This name comes from the 12 constellations the band contains. Most of these constellations have the names of animals, such as Aries (the Ram) and Taurus (the Bull).

Right: The zodiac constellation Taurus, the Bull. Ancient astronomers imagined they could see in this star pattern the head of a charging bull.

Written in the Stars

In ancient times, most people believed that the heavenly bodies somehow affected human lives. This idea is called astrology. Some people still believe in astrology today. Astrologers study the positions of the Sun and the planets in the constellations of the zodiac. They believe they can use this knowledge to predict events in people's lives.

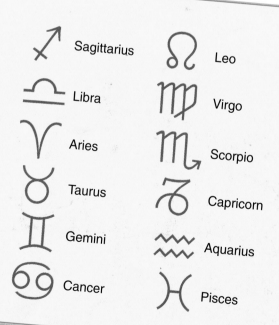

Sagittarius

Libra

Aries

Taurus

Gemini

Cancer

Leo

Virgo

Scorpio

Capricorn

Aquarius

Pisces

Comparing the Planets

The planets vary in all sorts of ways. Some are big, others small. Some are rocky, others are made of gas. Some spin slowly, others quickly. Some circle the Sun in days, others take years.

Our home planet, Earth, seems like a big place. But it is tiny compared with some of the other planets. The widest part of Earth is at the equator. If you drove a car around the equator at a speed of 50 miles (80 km) an hour, it would take you three weeks to circle Earth. But it would take you more than six years to drive a car around Jupiter!

We can compare the planets' sizes with the drawings on these pages. Jupiter is by far the biggest planet. Earth is the middle-sized planet. Four planets are bigger than Earth and four are smaller. Tiny Pluto is even smaller than Earth's Moon.

Jupiter may be huge for a planet, but it is tiny compared with the Sun. If the Sun were the size of a beach ball, Jupiter would be the size of a golf ball. Earth would be smaller than an orange seed!

Some of the moons in the solar system are bigger than some of the planets! Saturn's moon Titan is bigger than Mercury and Pluto. It measures 3,193 miles (5,140 km) across.

This picture shows all the planets drawn to the same scale. It shows just how small Earth and the other inner planets are compared with Jupiter and the other gas giants.

Mercury

Venus

Earth

Mars

Jupiter

Sun

Planet	Mercury	Venus	Earth	Mars	Jupiter	Saturn	Uranus	Neptune	Pluto
Diameter at equator (miles)	3,031	7,521	7,926	4,222	88,400	74,600	31,800	30,800	1,420
Average distance from Sun (million miles)	36	67	93	142	484	887	1,783	2,795	3,670
Rotates in:	59 days	243 days	24 hrs	24.6 hrs	9.9 hrs	10.6 hrs	17.2 hrs	16 hrs	6.3 days
Orbits Sun in:	88 days	225 days	365 days	687 days	12 yrs	30 yrs	84 yrs	165 yrs	248 yrs
No. of moons	0	0	1	2	16+	23+	15+	8	1

Among the gas giants, Saturn is slightly smaller than Jupiter. Uranus and Neptune are nearly the same size. Tiny Pluto is by far the smallest planet, smaller even than Earth's Moon.

Saturn

Uranus

Neptune

Pluto

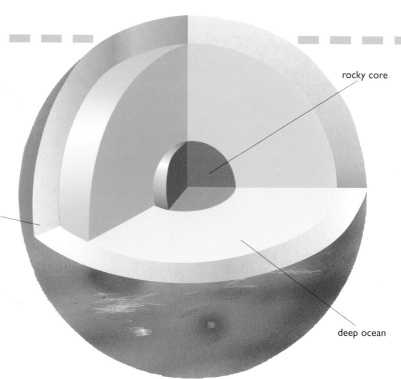

Right: A look inside Neptune, an example of a gas giant. It is made up of different layers, beginning with an atmosphere thousands of miles deep. We cannot see through this thick outer atmosphere. Underneath it is an even deeper ocean. And at the center is a rocky core.

rocky core

thick atmosphere

deep ocean

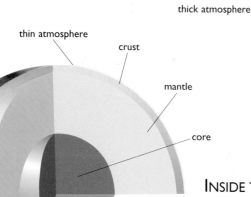

thin atmosphere

crust

mantle

core

A look inside Mars, an example of a rocky planet. Like Neptune, Mars is also made up of different layers. But it has only a very thin atmosphere, which we can see through. Underneath is a hard crust, which lies on top of the mantle. At the center is the core, which is probably solid, unlike Earth's liquid core.

INSIDE THE ROCKY PLANETS

In their structure, or makeup, planets fall into two main groups—rocky planets and gas giants. Earth is a typical rocky planet. At the surface is a hard outer layer of rock, called the crust. Earth is unique in that much of the crust is covered by water. Above the crust, a thin layer of gases forms the atmosphere.

Underneath Earth's crust is a thick layer of softer rock, called the mantle. This layer gets hotter the farther down you go. At Earth's center is a mass of metal, mainly iron. This is called the core. The temperature of the core is so high that the metal is liquid.

The planets Mercury, Venus, and Mars are made up of layers of rock in much the same way. That is why they are called the terrestrial, or Earth-like, planets.

INSIDE THE GAS GIANTS

The four giant planets, Jupiter, Saturn, Uranus, and Neptune, are quite different in makeup from the terrestrial planets. They are made up mainly of gases and cold liquid gases.

Jupiter is a typical gas giant. It has a very deep atmosphere of gases, mostly hydrogen gas. Underneath the atmosphere is a deep ocean of cold liquid gas. There is no solid surface at all. At the center of the planet is a small core of rock.

PLANET MOVEMENTS

All the planets move in two ways. Each planet travels in its orbit around the Sun. Earth takes a little over 365 days, or one year, to travel once around the Sun. Planets closer to the Sun take a shorter time to make one orbit because they have a shorter distance to travel. Planets farther from the Sun than Earth take a longer time because they have a longer distance to travel. Each planet also spins around on its axis. A planet's axis is an imaginary line that runs through it, around which it spins, or rotates. Earth takes 24 hours to spin around once. This is the period of time we call one day. Some planets spin faster than Earth; some spin more slowly. The table on page 21 shows how fast the planets spin and how long they take to orbit the Sun.

axis

North Pole

The Earth spins in space on an imaginary line called its axis. The line goes through the North and South Poles. All the planets spin around on an axis in a similar way.

South Pole

The Life Zone

There is one big difference between Earth and all the other planets. Earth is covered with living things. There are millions of different kinds of plants and animals, from tiny weeds to monster whales. Life does not exist on any other planet. One main reason for life on Earth is its position in the solar system. Its distance from the Sun means that it is not too hot and not too cold for life. We say that Earth lies in the Sun's life zone. Earth's neighbors in space, Venus and Mars, lie just outside the life zone. Venus is too hot, and Mars is too cold for life to exist.

The Hubble Space Telescope, photographed in orbit by astronauts in the space shuttle. Astronauts travel to the telescope every few years to check its instruments and other parts, such as the solar panels. These gold-colored panels on each side power the telescope. The masts sticking out from the telescope body carry radio antennae. These antennae receive signals from mission controllers and also send back the pictures the telescope takes.

Exploring the Solar System

Through the camera eyes of space satellites and probes, astronomers can take a close look at the planets and their moons, as well as asteroids and comets.

The ancient astronomers had no idea what the planets and the other bodies in the solar system were like. They knew of only five planets—Mercury, Venus, Mars, Jupiter, and Saturn. These planets appeared to them as bright shining stars.

In the 1600s, astronomers such as Galileo began using telescopes to look at the night sky. They discovered many new things about the solar system. For example, they learned that the planet Jupiter had moons circling around it.

Later, astronomers using more powerful telescopes began discovering new planets: Uranus (in 1781), Neptune (in 1846), and Pluto (in 1930). But even with the big telescopes available to modern astronomers we still cannot see many details on any of the planets. They are too far away.

The Hubble Space Telescope took this picture of Jupiter. It clearly shows the colorful bands of clouds in the atmosphere. It also shows Io, one of Jupiter's moons. The dark spot is Io's shadow.

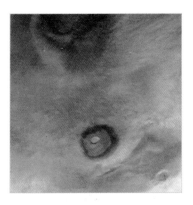

The surface of Mars, photographed by the space probe *Mariner 9* in July 1965. It was the first time that a probe had photographed a planet up close.

ROBOT EXPLORERS

Astronomers therefore use other ways of exploring the solar system. One way is by using robots. These robots are actually spacecraft that carry telescopes, cameras, and other instruments.

Telescopes on spacecraft get a much clearer view of the heavenly bodies than telescopes on Earth. Telescopes on Earth peer at the heavens through the atmosphere, which is full of dust and moisture that spoil the view. But spacecraft travel high above the atmosphere and are not affected by it.

ASTRONOMY SATELLITES

Two main kinds of spacecraft are used to explore space. One is the astronomy satellite, which stays in orbit around Earth and looks at the heavenly bodies from a distance.

The Hubble Space Telescope is an example of an astronomy satellite. It has been in orbit since 1990. And it is sending back amazing pictures of planets, comets, asteroids, nebulae, and stars. The telescope is a huge instrument, measuring 43 feet (13.1 m) long and weighing more than 11 tons. It circles in space about 320 miles (515 km) above Earth. It picks up light from the heavenly bodies with a curved mirror 95 inches (2.4 m) across.

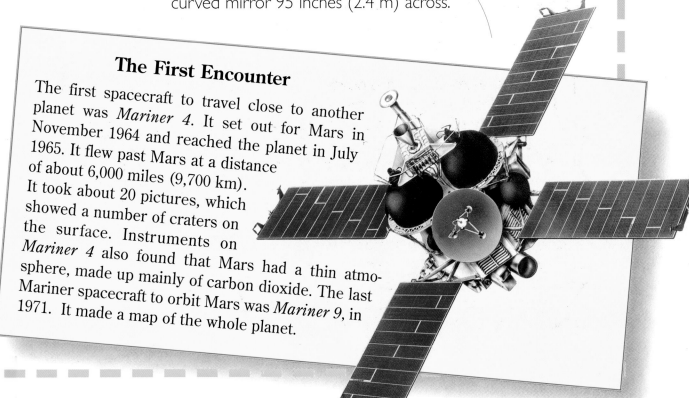

The First Encounter

The first spacecraft to travel close to another planet was *Mariner 4*. It set out for Mars in November 1964 and reached the planet in July 1965. It flew past Mars at a distance of about 6,000 miles (9,700 km). It took about 20 pictures, which showed a number of craters on the surface. Instruments on *Mariner 4* also found that Mars had a thin atmosphere, made up mainly of carbon dioxide. The last Mariner spacecraft to orbit Mars was *Mariner 9*, in 1971. It made a map of the whole planet.

August 20, 1977: the launch of *Voyager 2*.

PROBES TO THE PLANETS

Astronomers also use a spacecraft called a space probe to explore the solar system. Space probes are unmanned vehicles that leave Earth behind and travel into the depths of space. Probes are sent to the Moon, to the planets and their moons, and to comets and asteroids.

The first successful probes to the planets were sent to Mars and Venus in the 1960s. Since then, probes have visited all the planets except Pluto. They have sent back amazing information and pictures. They have shown us that Mercury looks much like the Moon; that Jupiter's moon Io has volcanoes; that winds on Saturn can blow at more than 1,000 miles (1,600 km) an hour; and that Neptune's moon Triton has geysers.

FAST MOVERS

Very powerful rockets are needed to launch a probe to a target that may be billions of miles away. A rocket must make a probe go fast enough to escape the pull of Earth's gravity. To escape, a probe has to be launched at a speed of 7 miles (11 km) per second. At this speed, you could fly across the Atlantic Ocean in about 8 minutes!

Voyager 2 set off on its historic journey to the outer planets atop a Titan-Centaur rocket in August 1977. Less than two years later, it was flying past the giant planet Jupiter. Boosted by Jupiter's gravity, it sped on to Saturn, then to Uranus, and finally to Neptune. It arrived at Neptune after spending 12 years traveling through space.

Jupiter: July 9, 1979

Saturn: August 25, 1981

Highlight Missions

Probe	Month launched	Target
Luna 2	September 1959	Moon
Mariner 4	November 1964	Mars
Surveyor 1	May 1966	Moon (lander)
Venera 4	June 1967	Venus
Lunokhod 1	November 1970	Moon (rover)
Mariner 9	May 1971	Mars (orbiter)
Pioneer 10	March 1972	Jupiter
Pioneer 11	April 1973	Jupiter, Saturn
Mariner 10	November 1973	Mercury
Viking 1	August 1975	Mars (orbiter, lander)
Voyager 2	August 1977	Jupiter, Saturn, Uranus, Neptune
Magellan	May 1989	Venus (orbiter)
Galileo	October 1989	Jupiter
Ulysses	October 1990	Sun
Sojourner	December 1996	Mars (rover)
Cassini	October 1997	Saturn's moon Titan

Left: Triton, Neptune's largest moon, was *Voyager 2*'s last port of call. This icy moon is covered with a pinkish snow of frozen gases. Dark material thrown out from icy volcanoes is visible in places.

FLY-BY, ORBIT, OR LANDING

Most probes carry out a fly-by, which means they fly past a planet or other body. The *Voyager* probes to the outer planets and the *Giotto* probe to Halley's comet carried out fly-bys. Some probes fly to a planet and then go into orbit around it. The *Magellan* probe to Venus and the *Galileo* probe to Jupiter were orbiters.

Other probes travel to a planet or a moon and drop landing craft down to the surface. The *Viking* probes to Mars did this. The *Pathfinder* probe that landed on Mars carried a wheeled vehicle down to the surface. This rover, called *Sojourner*, traveled around investigating nearby rocks. It was guided by scientists back on Earth.

Voyager 2

Neptune:
August 24, 1989

Uranus:
January 24, 1986

Other Solar Systems

Out in space, in other solar systems, there may be planets like Earth. These planets might swarm with life like Earth does.

The Sun was born in one of the great clouds of gas and dust that exist in space. As far as we know, all stars form in the same way. This means that other stars could form some kind of solar system, containing planets like those in our own solar system.

Some astronomers think they can detect a few stars with planets. They notice that these stars wobble slightly, as though they are being pulled by unseen bodies, which could be other planets. The planets themselves would be too small to be seen even in the most powerful telescopes on Earth.

Above: Astronomers have found a disk of matter around a star named Beta Pictoris. Over time, the matter in the disk could lump together to form planets.

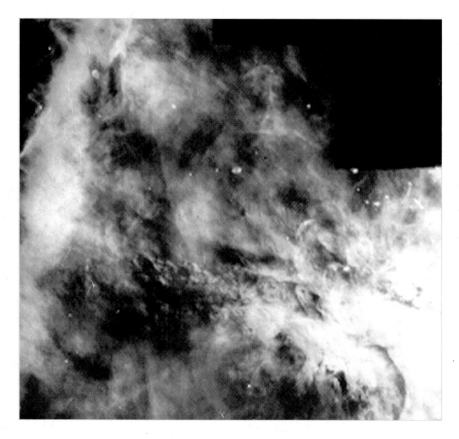

Right: This is part of the famous Orion nebula, as seen by the Hubble Space Telescope. It shows a number of small disks where solar systems are forming. Before the Hubble pictures, astronomers had never seen such disks clearly. They call these newly born solar systems proplyds (short for protoplanetary disks).

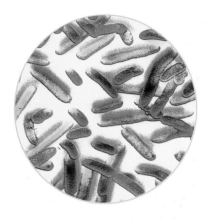

Bacteria like this one, E. coli, are found everywhere on Earth, in the air, in the ground, and in other living things. Scientists think that similar kinds of organisms were among the first forms of life on Earth.

Even using space telescopes, we cannot see individual planets near distant stars. But we are able to spot disks of matter around some of the stars. The Hubble Space Telescope has sent back spectacular pictures of such disks in the Orion nebula. One day, the gas and dust in these disks could lump together to form planets, which is what happened in our own solar system long, long ago.

PLANETS LIKE EARTH

In our Universe, there are many stars like the Sun. And many of these stars are probably the center of a solar system. And in many of these solar systems, there are probably planets like Earth. Because there are billions of stars like the Sun, there could be billions of planets like Earth.

Many of these planets might circle in the life zones of stars, where the temperature would be not too hot and not too cold. So on these planets, life of some kind could exist. There could even be intelligent life-forms like human beings.

How Life Began

Scientists do not know whether life exists elsewhere in space. One reason for this is that they do not really know how life came about on Earth. Many scientists believe that life began with chemicals that formed in Earth's atmosphere billions of years ago. They formed when lightning acted on the gases it contained. The chemicals rained into the oceans and joined to make very simple living things. These early life-forms were like the tiny life forms we know as bacteria. Then over many hundreds of millions of years, more advanced life-forms—plants and animals—developed.

Above: The *Voyager* space probes are carrying copies of a record disk called Sounds of Earth, just in case alien beings come across them one day. On the disk are messages from Earth, and coded pictures and sounds from our planet.

Below: In 1974, astronomers sent a coded message (below) into space from the powerful radio telescope at Arecibo in Puerto Rico (below right). The radio message is now traveling among the stars, carrying information about where our planet is and the life-forms that live on it.

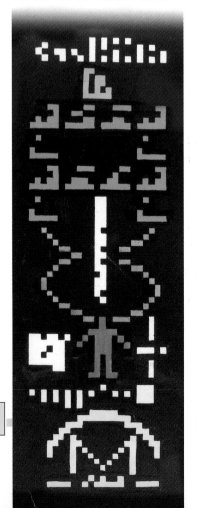

SEARCHING FOR LIFE

We use the terms alien and extraterrestrial to refer to life that might exist elsewhere in space. The word *extraterrestrial* means "from outside Earth." We do not know for certain if there are any aliens out there. We cannot travel to planets in other solar systems to look for life because they are too far away. And as far as we know, other alien life-forms cannot travel to visit us for the same reason.

Some astronomers are using radio telescopes to try to make contact with alien life-forms in space. The big dishes of these telescopes can pick up the faint waves from distant stars. For years astronomers have been tuning in to radio waves that come from outer space. They listen for signals that might contain messages from intelligent beings. This work is known as SETI, or the search for extraterrestrial intelligence.

No intelligent signals have been received yet, but the astronomers are still listening. Maybe one day they will pick up messages from distant planets, and we will then know for certain that we are not alone in the Universe.

Glossary

alien: something that is not from Earth, as in alien life

asteroid: a small rocky body that orbits between Mars and Jupiter

astrology: a belief that the positions of stars and planets somehow affect people's lives

astronomy: the scientific study of space and the bodies in it

atmosphere: the layer of gases around Earth or another heavenly body

comet: a small lump of dust and ice that shines when it nears the Sun

constellation: a group of stars that forms a pattern or shape

crater: a pit in the surface of a planet or moon

crust: the hard surface of a rocky planet like Earth

extraterrestrial: from outside Earth

fly-by: a type of voyage carried out by a space probe, in which it flies by a planet or other heavenly body

gravity: the attraction, or pull, that a heavenly body has on objects on or near it

heavenly body: an object, such as a moon or a planet, that appears in the sky

mantle: a layer of rock underneath the crust of a rocky planet

meteor: a lump of rock that forms a streak of light produced when it burns up in Earth's atmosphere

meteorite: a lump of rock from space that hits a planet or moon

meteoroid: a small lump of rock or metal that orbits the Sun

moon: a natural satellite of a planet

nebula: a cloud of dust and gas in space

orbiter: a space probe that goes into orbit around a planet or moon

planet: a large heavenly body that orbits the Sun

probe: a spacecraft that travels from Earth to a heavenly body

radio telescope: a telescope with a large dish-shaped antenna that collects radio waves from space

satellite: an object that circles in orbit around a larger body. Space satellites are spacecraft that orbit Earth.

solar: having to do with the Sun

solar system: the family of bodies that orbit the Sun, including Earth, the other planets, and their moons

star: the most common heavenly body, a huge ball of hot gases that gives out enormous energy, mainly in the form of heat and light

terrestrial: like Earth, or having to do with Earth

Universe: space and everything in it, including stars, planets, moons, gas, and dust

zodiac: an imaginary band in the heavens in which the planets are found

Index